HAYNER PUBLIC LIBRARY DISTRICT-ALTON
W9-ACB-466

Discovering WORMS

Jennifer Coldrey

The Bookwright Press
New York · 1986

Discovering Nature

Discovering Bees and Wasps
Discovering Snakes and Lizards
Discovering Spiders
Discovering Worms

Further titles are in preparation

All photographs from Oxford Scientific Films

First published in the United States in 1986 by
The Bookwright Press
387 Park Avenue South
New York, NY 10016

First published in 1985 by
Wayland (Publishers) Limited
61 Western Road, Hove
East Sussex BN3 1JD, England

Second impression 1986

ISBN 0-531-18046-8
Library of Congress Catalog Card Number: 85-62087

Typeset by Planagraphic Typesetters Limited
Printed in Italy by G. Canale & C.S.p.A., Turin

Contents

The World of Worms

Everyone knows what a worm is like — long and thin, pink and wriggly, and soft and slimy to the touch. But not all worms are like this. There are thousands of different kinds of worms in the world. Some are thin and threadlike, others flat and leaf-shaped; some are covered with scales or bristles, some have tentacles while others have suckers. No true worm has a backbone.

The colorful sea mouse is actually a bristleworm although it looks more like a small, furry mouse.

These tube worms, on a coral reef, use their brightly colored tentacles to catch food.

Different worms live in different places. Many burrow into the soil or live under rotting leaves and compost. Others live underwater in ponds, streams and ditches. There are also vast numbers of fascinating and often brightly-colored worms living in the sea.

Many worms live in burrows or tubes where they can hide away from their enemies. An earthworm's burrow is also useful in preventing its body from drying out, for worms cannot survive in the open air for very long. Their soft bodies are covered with very thin skin through which they breathe. They need to keep their bodies moist to allow the oxygen to pass in through the skin.

There are many strange and unfamiliar worms that are not really related to true worms at all. Some are hardly ever seen because they live as **parasites** inside the bodies of other animals.

What Worms Look Like

Earthworms

Earthworms belong to a large and important group of worms called **annelids.** They are also known as segmented worms because their bodies are divided into many rings or segments. All true worms are annelids. An earthworm's body is long, thin and cylindrical in shape and it tapers to a point at each end. The mouth is at the front end, but there is no

You can clearly see this earthworm's segmented body and its girdlelike clitellum.

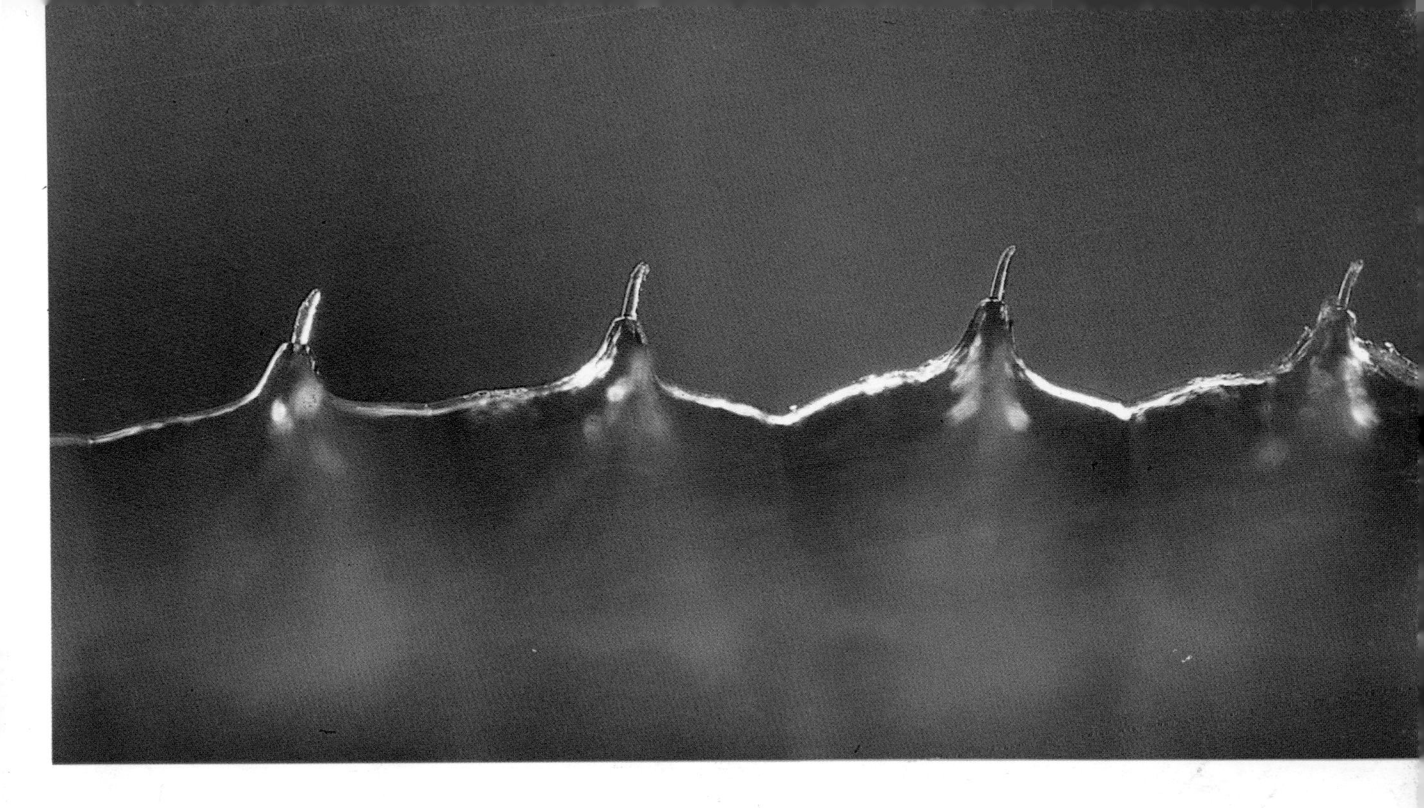

A close-up of an earthworm's bristles.

real head, and an earthworm has no eyes, ears or nose.

The earthworm's pinkish-brown body is covered with a thin, slimy skin which is sensitive to touch and also to light. The skin is not entirely smooth because on every segment, except for the first and last, there are four pairs of tiny bristles sticking out from the underside of the body. These bristles cling to the soil and help the earthworm when it moves.

A fully grown worm often has a swollen band or girdle (the **clitellum**) around its body, which plays a special part in breeding.

There may be as many as 3,000 different kinds of earthworms in the world. Many are only a few centimeters long, but the largest, the giant Australian worm, is over 3 meters (10 feet) long.

Bristleworms

Bristleworms live in the sea. Like earthworms, the bodies of bristleworms are divided into segments, but they have many more bristles than earthworms do. Their stiff bristles usually stick out in tufts from paddlelike flaps on the sides of their bodies. These flaps are found in pairs on each segment. In some bristleworms the flaps are very small. In others they are

Below *Two brightly colored bristleworms found on a sandy beach in Australia.*

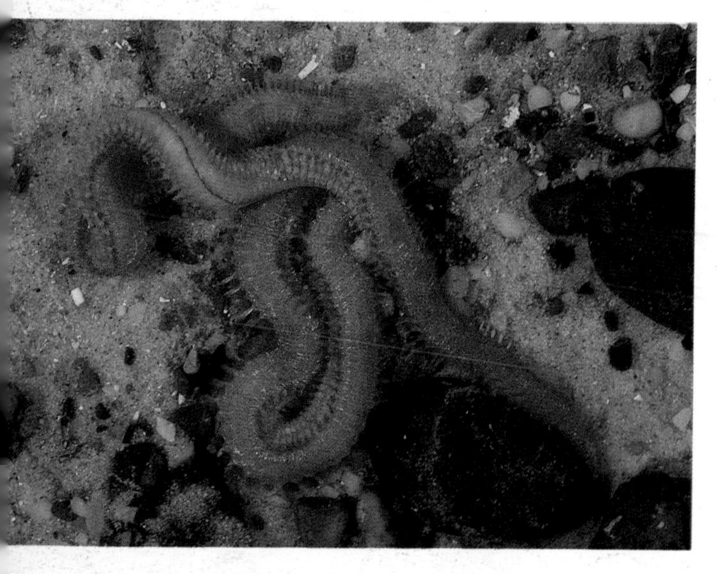

Above *This tube-dwelling bristleworm has a mass of red gills and many long tentacles.*

large and leaflike and are useful for swimming in the sea and for crawling on the sea-bed.

Most bristleworms have much more definite heads than earthworms do. Many have several small eyes as well as various feelers and tentacles. These are used for seeing, touching and smelling, and help the worm to find and catch its food. The bristleworm often has a snout or **proboscis** which it uses for feeding as well as for digging into the sand. Those that eat other animals usually have well-developed mouths with jaws and sometimes teeth.

Below *The bright red gills of this bristleworm are used for breathing.*

Many bristleworms have bright red **gills** sticking out from their bodies. These are well supplied with blood to absorb oxygen for breathing.

There are hundreds of different kinds of bristleworms. Some live permanently in tubes or burrows on the sea-bed and all we can see of them is a mass of gills or a crown of tentacles sticking out into the water.

Leeches

Most leeches live in freshwater ponds and streams. They are also found on land in warm, wet parts of the world, and a few live in the sea as parasites on fish. Leeches feed by eating other animals or sucking their blood.

Leeches' bodies are divided into segments, and leeches are closely related to earthworms and bristleworms. However, they are different in several important ways. Leeches all have two suckers on their bodies. The one at the back end is the bigger while the smaller one surrounds the mouth. Their suckers are used for clinging to stones and plants as well as to their **prey.**

Leeches use their suckers to attach themselves to plants and stones as well as to cling to their prey.

This horse leech moves through the water with a graceful wavelike motion.

Leeches have no bristles on their bodies. They have small heads with one or several pairs of small eyes. Some have mouths with jaws and teeth, while others have proboscises for killing and devouring their prey.

Leeches have much flatter bodies than most other worms. When resting they tend to shrink into an oval leaflike shape, but they can change dramatically by moving and stretching their bodies in various ways. When crawling, they use their suckers as feet and move along by "looping" their bodies, as some caterpillars do. Leeches can swim well too.

Where Worms Live

On Land

Some worms live under stones or logs and among rotting leaves or compost, but most earthworms live in burrows underground. They prefer the soft, rich, undisturbed ground under lawns, meadows and other grasslands. Earthworms come out of their burrows only on warm, damp nights to look for food or perhaps to mate. When out on the surface they keep their tails safely anchored inside their burrows and shoot back quickly into the ground if danger threatens.

When burrowing into the soil, an earthworm pushes its pointed snout into the ground. It then grips with its bristles and, by alternately tightening and relaxing the muscles in different parts of its body, it gradually works its way down.

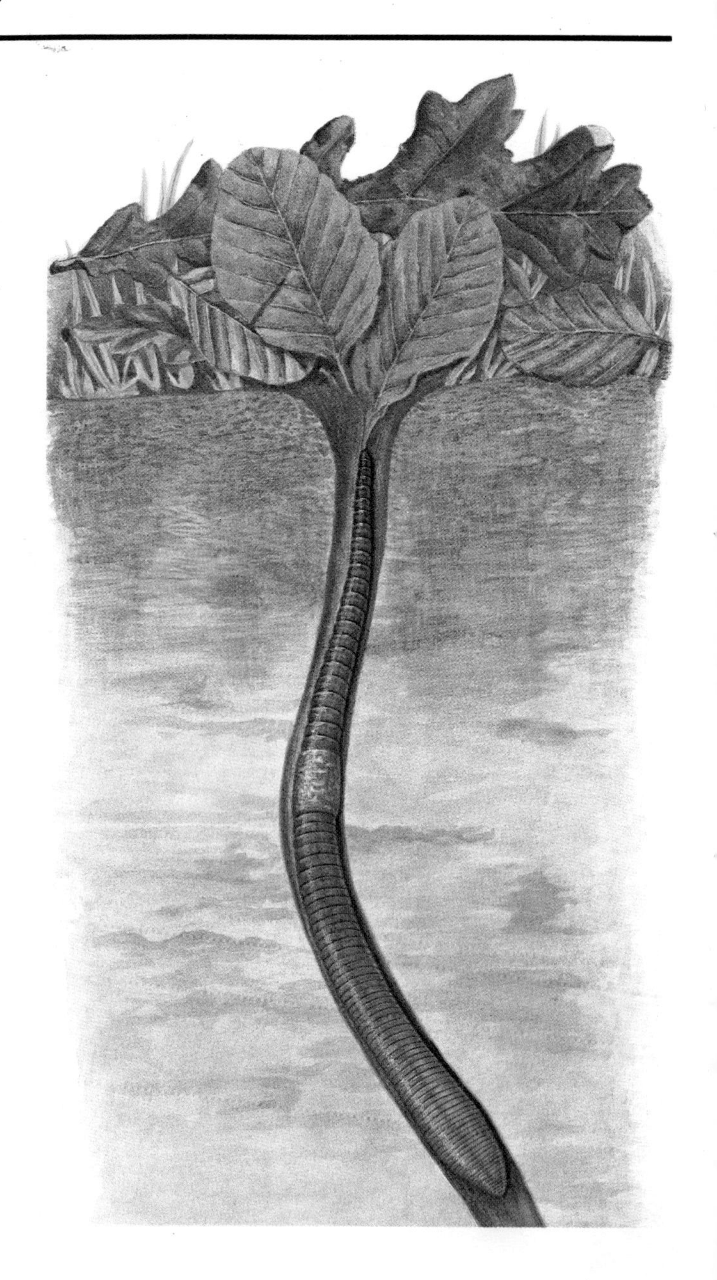

A worm pulling leaves into its burrow.

Some earthworms make their castings on the surface of the lawn.

Earthworms also make their burrows by eating their way through the soil, which passes through their bodies and comes out as castings. You can often spot worm castings on a lawn.

Most worms plug the entrances to their burrows with leaves, sticks or even pebbles, mixed together with mud. This protects them from the cold and rain as well as from drying out in hot weather. Earthworms are very useful in the soil. They act like tiny plows, mixing the soil around and breaking it up into fine pieces. Their burrows also let in air and help to drain the ground.

In Fresh Water

Many freshwater worms are closely related to earthworms.

Freshwater worms are easy to find in the shallow water of ponds and ditches, as well as at the edges of rivers and streams. Some of them crawl over stones and waterweeds. Others burrow into the mud at the bottom, while some very tiny worms live inside snail shells or in the cases of insect **larvae.**

The burrowers often line their tunnels with mud and silt. Some build tubes that stick up slightly into the water. Tubifex worms are some of the commonest tube-dwellers. They live in groups and are found in thick mud, and sometimes in deep or polluted water, as well as in the shallows. There is often

Tubifex worms wave their tails around in the water. This stirs up the oxygen so that more can be absorbed through the skin.

very little oxygen where they live, but tubifex worms can survive because they have plenty of bright red blood which absorbs oxygen very easily. They live upside down in their tubes, with their heads buried in the mud to find food, and their tails waving gently to and fro.

Many other freshwater worms are bright red in color. Like tubifex worms, their rich blood supply helps them to breathe. Some freshwater worms are

These red pond worms have transparent body walls through which you can see their blood.

colorless and transparent, and it is very difficult to see them in the water. This helps to protect them from enemies, such as fish. Many leeches live in fresh water, and so do hairworms, roundworms and flatworms, though these last three types of worms are not true worms.

In the Sea

Most of the worms living in the sea are bristleworms. A few swim about freely in the water, using their paddlelike limbs to push themselves along. Many bristle-worms live along the seashore and in estuaries where they crawl over rocks and seaweeds as well as burrowing into the sand or mud. Some of these worms can swim well too.

A lugworm in its burrow. It feeds by swallowing sand from which it filters out the goodness.

Bristleworms have many enemies and those that do not burrow often hide among seaweed, under stones or in rock crevices. Some bristleworms spend their whole lives in a burrow. The lugworm is one of these.

Lugworms live between the tides on sandy shores. They build U-shaped bur-rows lined with slime. They feed on sand and other small particles that filter down into the front end of the burrow. On the surface of the sand a small pit marks this end of the burrow. At the other end the lugworm pushes out its castings.

When the tide is in, the lugworm uses its paddles, as well as rhythmical movements of its body, to draw water down into its burrow. Oxygen is absor-bed by several pairs of feathery, red gills which stick out along the middle of its

body. Many other burrowing worms breathe and also feed by drawing water through their burrows in this way.

You can often find the large coiled castings of lugworms covering vast areas of a beach at low tide.

It is difficult to see the sand mason's body inside its tube, but sticking up from its head is a clump of long tentacles which it uses to catch food.

Many bristleworms build their own tubes to live in. Some make them out of sand, and you can sometimes see their tubes sticking up on the beach at low tide. One of these is the sand mason. It builds a rather ragged tube made of rough sand and tiny pieces of shell bound together with sticky **mucus.**

Peacock worms, or fan worms, also live in tubes of sand. They have beautiful crowns of tentacles on their heads. The tentacles open out like fans when the tide covers them. Like the sand masons, fan worms use their tentacles to catch tiny particles of food from the water.

Some fan worms protect themselves by living in thick, tough tubes of jelly. Others build hard, white tubes of lime in which to live. You can often find these attached to seaweed or stones, as well as on the shells of animals like mussels, clams and crabs. Some look like little white snails; others grow tall and trumpet-shaped, while some are in coiled or twisted groups, and look very much like a mass of cooked spaghetti.

Tube worms cling to the insides of their tubes with the bristles on their bodies. If danger threatens they quickly pull in their tentacles and hide inside their protective homes.

Many lime-tube dwellers have one tentacle shaped like a plug, which blocks the end of the tube when the worm retreats into its home.

Food and Feeding

Earth-eaters and Sand-eaters

Many burrowing worms eat the soil, sand or mud into which they tunnel. The soil that earthworms swallow contains plenty of food in the form of dead plant and animal remains, bacteria and various minerals. This passes down the gut and into the part of the stomach called the **gizzard** where it is ground into fine pieces. The soil then goes along the **intestine** where the food is absorbed into

The fleshy lobe at the tip of an earthworm's snout probes around and helps it to find and grasp food.

the body. The remains of the soil pass out as a worm casting.

Earthworms also feed on leaves and other plant remains which they collect on the surface of the ground by reaching out from their burrows. They often pull leaves down into their burrows and nibble the ends. By forming their castings at the surface, earthworms make the soil rich and fertile. The leaves they drag down also add to the goodness of the soil.

Many of the worms living underwater feed in the same way as earthworms.

The peacock worm uses its crown of tentacles to catch tiny particles of food in the water around it.

There are plenty of microscopic creatures, as well as the rotting remains of plants and animals, in the mud and silt lying on the bottom. Some underwater worms have proboscises which they push out from their mouths to suck in food. Many burrowing bristleworms have tentacles which stick out from the sand or mud and pick up food from the surface.

Flesh-eaters and Bloodsuckers

Many of the worms that crawl or swim around freely in the sea are **carnivorous.** They are active hunters and feed by killing and eating other animals. Many bristleworms have long proboscises armed with teeth and jaws, which they use to attack and eat their victims. They feed on a variety of small creatures, including tiny fish, snails, and other worms. Their eyes, feelers and tentacles help them to find their prey. But not all bristleworms are active **predators** — some use their

The head of a predatory bristleworm is actually about 5 mm (¼ inch) long.

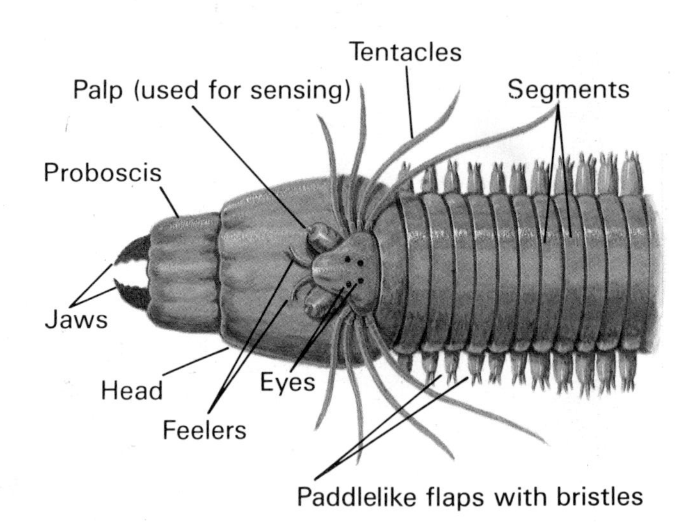

Below *Ragworms will eat dead animals but they also hunt and kill live prey.*

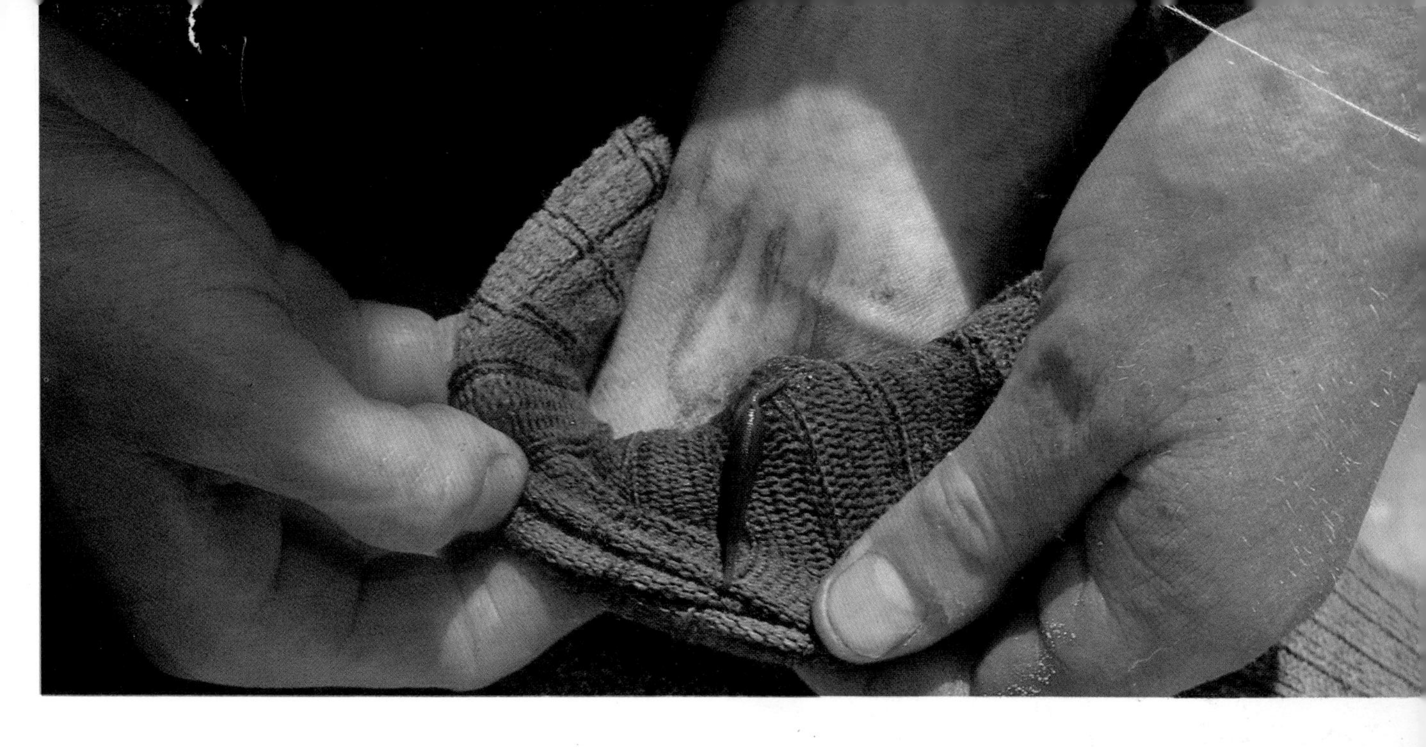

Bloodsucking leeches produce a substance in their saliva which keeps the blood of their victims from clotting.

rasping teeth to scrape seaweeds, sponges and small creatures off the rocks, while others feed mainly on dead animals.

Some hunting worms use the powerful muscles of the mouth and proboscis to suck out the insides of their prey. Many leeches feed in this way, attacking small animals. Other leeches are bloodsuckers. They attach themselves firmly by their suckers, then pierce through the skin of the victim and suck its blood. Fish leeches use a proboscis for this, but the larger leeches have jaws and teeth that cut into the tougher skins of animals such as frogs, birds and humans.

Bloodsucking leeches do not usually kill their prey. They feed as parasites. Some stay permanently attached to one **host,** but others drop off and move to a different host each time they want a meal.

Reproduction

A bristleworm shedding its eggs into the sea.

Laying Eggs and Regeneration

Different types of worms produce young in different ways. Most worms lay eggs. The eggs need to be fertilized by **sperm** from a male before new, young worms can start to grow. Laying eggs in the water is called spawning.

Some bristleworms spawn on the sea-bed and even in their own burrows, but most bristleworms shed their eggs and sperm into the sea. The sperm swim toward the eggs and fertilize them. Males and females need to be close together when this happens.

The fertilized eggs hatch out into tiny larvae, which swim around for several weeks feeding on minute creatures in the sea. The larvae are transparent, with many hairs or bristles that help to keep them afloat. Although they are difficult to see, they are eaten in enormous

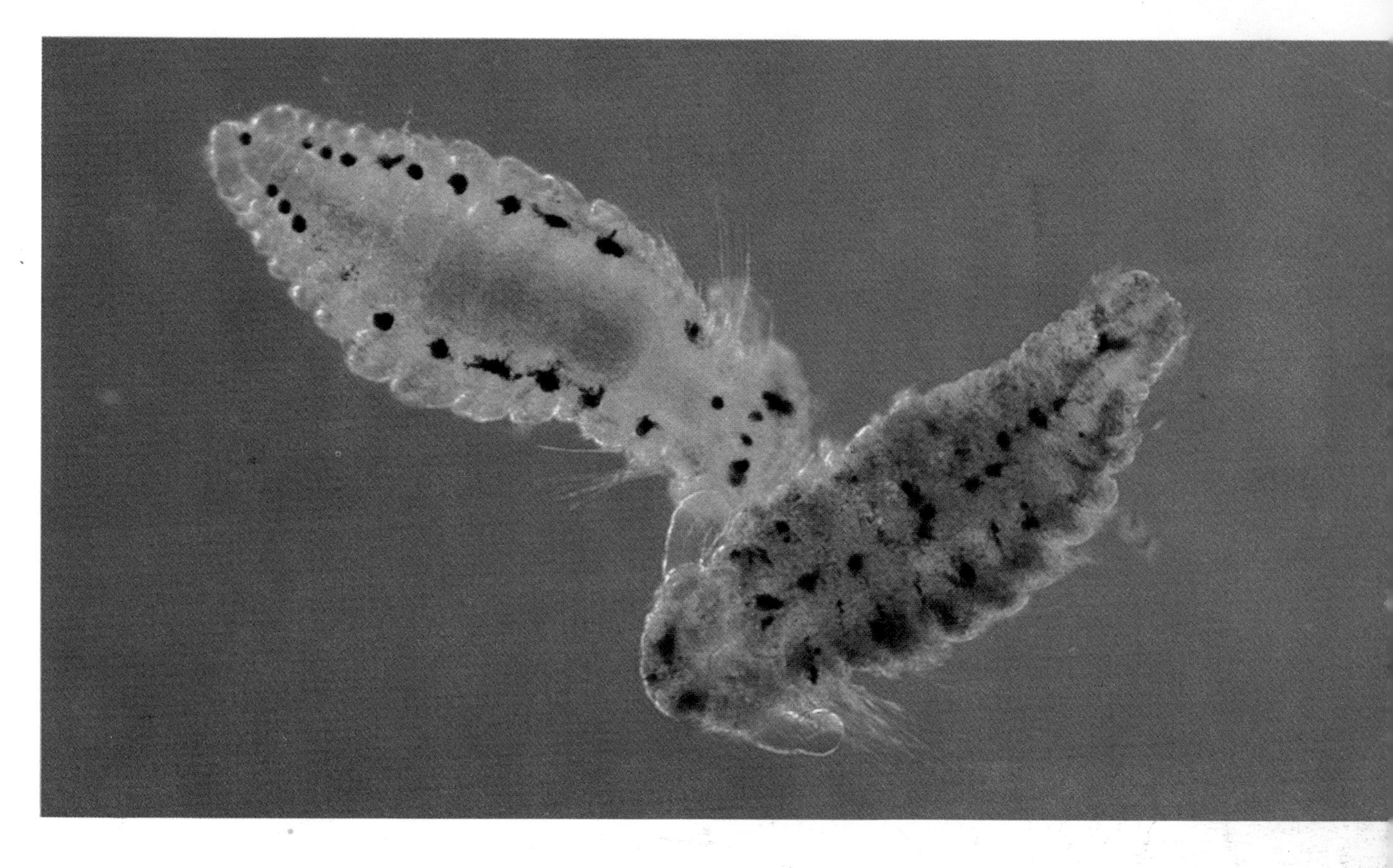

numbers by fish and other predators. The few that survive eventually sink to the bottom where they grow into worms. Not all bristleworm larvae swim around freely in the sea. Somc develop in the mud on the bottom, others inside the tubes or burrows of their parents.

Some worms can split their bodies into pieces. Each piece may then grow into a new worm. This is called regeneration. It happens to many worms when they are damaged by accident, but it is also used regularly by some freshwater and marine worms as a way of multiplying.

These bristleworm larvae are only 1 to 2 mm long. You can see the tiny hairs that help keep the larvae afloat.

Hermaphrodites

Earthworms and leeches are hermaphrodites, which means that each animal is both male and female and can produce both eggs and sperm.

When they join together or mate, they fertilize each other's eggs. Most earthworms mate in their burrows, but the common earthworm comes to the surface, and usually mates at night. When mating, the two worms lie with their bellies touching and their heads facing in opposite directions. Each worm then passes its sperm into a special pouch on the other worm's body. The worms then separate and return to their burrows.

When the worm is ready to lay eggs, the clitellum swells up and forms a sticky tube of mucus. This moves forward along the worm's body, first picking up four or five eggs from the female opening

When mating, earthworms produce a lot of slimy mucus which holds them tightly together.

and then the other worm's sperm which fertilize the eggs. The earthworm wriggles backward out of the mucus tube which hardens and seals around the eggs to form an oval shell called a cocoon. The brown cocoons are left in the soil and after several weeks one or two fully formed young worms hatch out. The other eggs usually die, providing food for the ones that survive.

Leeches mate in the same way and leave their cocoons attached to stones or plants under water. Some leeches carry their cocoons around with them. Others stay nearby and guard their eggs until they hatch.

This earthworm cocoon is about the same size as a grain of wheat.

Below *Some leeches carry their young around for several weeks on the undersides of their bodies.*

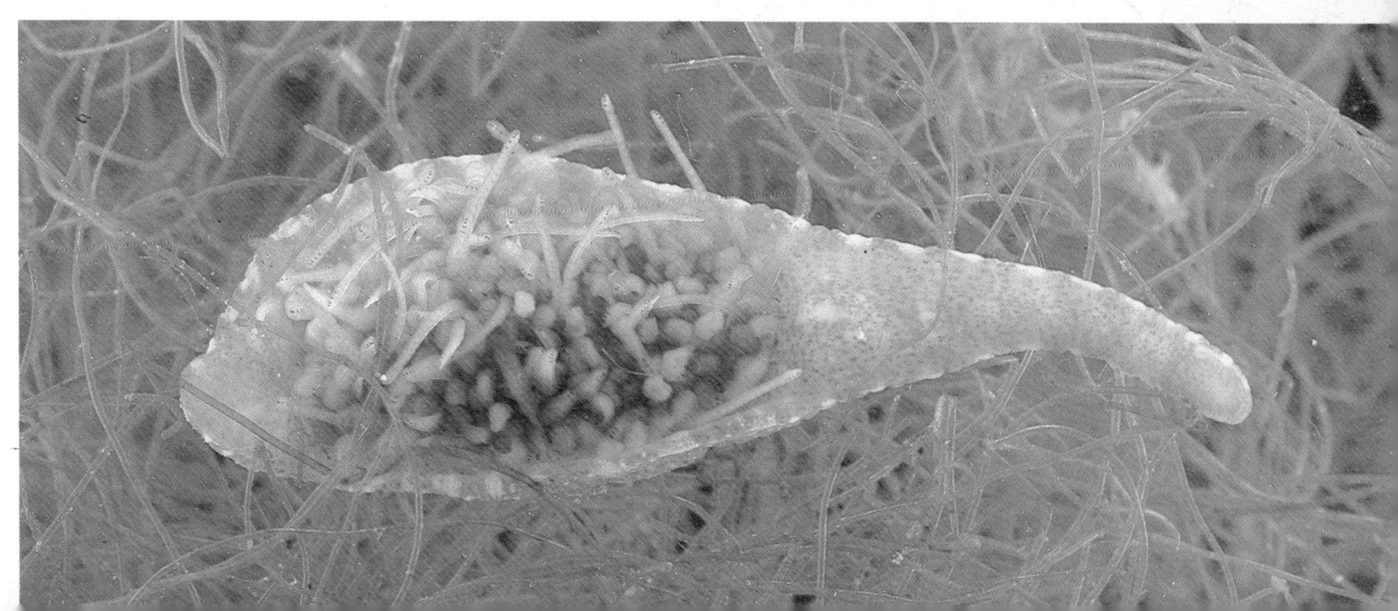

Survival in a Dangerous World

Enemies

Worms have many enemies. They are soft and juicy and make a tasty meal for many animals. Earthworms are eaten by moles, badgers, shrews, frogs and toads, as well as by many different birds.

A toad makes a meal of an earthworm.

Most of these animals hunt at night when earthworms are out on the surface of the ground.

But birds catch worms during the day, even when they are down their burrows. You have probably seen a robin or blackbird with its head cocked on one side, listening for noises underground. If it hears a worm tunneling below, it stabs its beak into the earth and grabs its victim. Birds often have to struggle to pull out worms, because they cling tightly to the soil with their bristles. Sometimes their bodies snap in two and the bird only gets half a worm.

Moles are the chief enemy underground. They feed on almost nothing but worms, and often store large numbers of them in special larders. They bite off the front ends of the worms to prevent them from escaping. Other creatures prey on earthworms under the ground, including

centipedes, and some beetles and slugs.

Worms living under water are hunted mainly by fish, as well as by water birds. Those living in estuaries and along the seashore are eaten by various shore birds, which probe into the mud or sand with their long, pointed bills.

Above *This robin has caught a worm to feed to its young.*

When burrowing through the soil, earthworms often fall into moles' tunnels where they are caught and eaten.

Man and Other Dangers

People sometimes kill or harm worms too. Fishermen collect them for bait and earthworms are often poisoned by farmers and gardeners using weed-killers and insecticide sprays to protect their crops. In some parts of the world people even eat earthworms because they are a rich source of **protein.**

Sometimes earthworms can spread diseases through the ground. They are also a nuisance when they leave their castings on lawns, golf courses and bowling greens. People then try to get rid of them by treating the grass with poisonous chemicals.

On sandy shores fishermen dig up lugworms and ragworms for bait.

If an earthworm is chopped in half by a gardener's spade, the front end will eventually grow a new tail.

Human beings also cause harm to worms (and many other creatures) living underwater, when they pollute lakes, rivers and the ocean with oil and poisonous materials.

Worms cannot survive in very hot or very cold weather. Under these conditions many burrowing worms dig deeper to escape. In very hot, dry weather, many worms coil themselves into a tight ball and go into a resting period. A protective cocoon forms around their bodies and they do not eat or move until it rains.

In spite of the many dangers all around them, worms have amazing powers of survival. They are able to repair and even regrow parts of their bodies that become damaged. Some can even produce several complete new worms when they are torn into pieces.

Some Strange and Different Worms

Ribbon Worms and Horsehair Worms

There are many other worms that are not closely related to segmented worms. Some have simple bodies with no segments or bristles, and many are named after their shapes. The ribbon worm is long and ribbonlike, with a slightly flattened body and a snakelike head with several small eyes. You find ribbon worms on the seashore where they hide under stones or in the mud and sand. Some are very long — bootlace worms have been found up to 10 meters (33 feet) in length.

This ribbon worm lives in the warm water off Bermuda.

A mass of horsehair worms, so named because people used to think that hairs from a horse's tail had dropped into the water and come to life!

The ribbon worm is carnivorous and often eats other worms. It has a long, threadlike proboscis, which shoots out rapidly from its mouth and captures the prey by coiling around it.

Horsehair worms live mainly in fresh water or in damp soil. They can be anything from 10 cm to 1 meter (4 inches to 3 feet) long, and have very thin bodies, only 1 or 2 mm thick, about as thick as your finger nail. There is no obvious head or mouth. These threadlike worms, usually brown or black, are often found tangled together in a writhing mass.

Horsehair worms have a very strange life history. Their eggs hatch into larvae, which are eaten by various animals. They only survive as parasites inside the bodies of certain insects, where they grow into worms and finally break out when the insect host is near water.

Flatworms

It is easy to see how flatworms get their name. They are flat, sometimes leaf-shaped, and often look more like slugs than worms. The flatworm has a simple body. It has a mouth, but no **anus,** so all its food-remains come out of the mouth too. It is sometimes difficult to tell which end is which, but the head usually has several eyes, and sometimes tentacles.

This strange, slimy creature is a tropical land-living flatworm.

Flatworms usually live in water, although some can survive on land in warm, damp places. Many live in the sea, where they creep along on the bottom, as well as occasionally swimming. Flatworms can move and change shape by flexing the muscles in their bodies. They also move by gliding along on a cushion of tiny, beating threads, called cilia. Flatworms have very slimy skin. This helps them to cling to stones and weeds as they move.

Some flatworms feed on dead animals and plants, but most are predatory. They eat small animals, such as worms and young insects. The flatworm pushes a muscular tube from its mouth, which latches onto the prey and sucks in its body.

Flatworms can grow new body parts very easily if they are damaged. Many flatworms can **reproduce** by splitting themselves into two or more pieces. Each of these pieces then grows into a new flatworm.

This freshwater flatworm has just finished feeding through the tube sticking out from underneath its body.

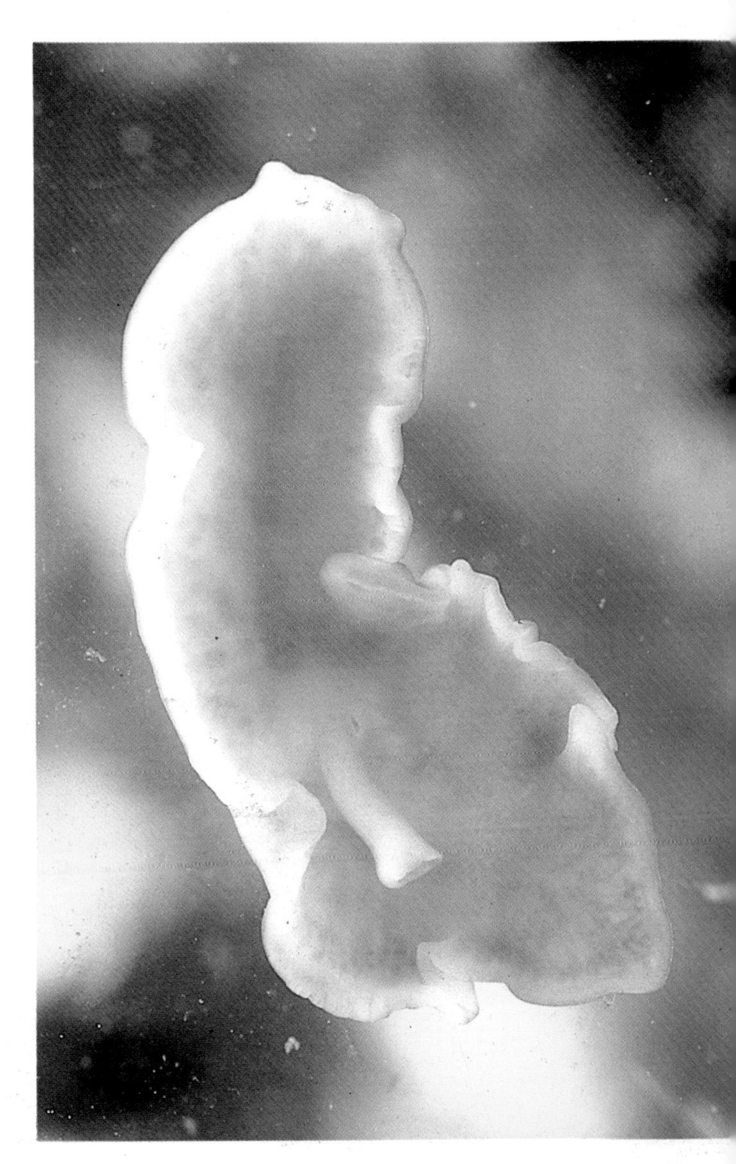

Parasitic Worms

Many worms live as parasites inside other animals or plants. Tapeworms live in the intestines of larger animals such as birds, fish and humans. They cling to their hosts with their tiny heads which are usually armed with hooks and suckers. Their ribbonlike bodies are made up of long strings of identical sections. Some

All these tapeworms were found inside this dead fish. Some tapeworms can grow as long as 9 meters (30 feet).

tapeworms grow new sections all the time, becoming longer as they get older.

Tapeworms have no mouths or guts. They feed by absorbing liquid food through the skin from the intestines of their hosts. Their ripe eggs pass out in the host's droppings, but they can only hatch if they are taken into the body of another animal. They finally become adult tapeworms when they are swallowed by the right host.

Roundworms are common parasites in many animals including humans. They may live in various parts of the body, such as the gut, lungs, kidneys, muscles, or even inside the eye. The eggs are often eaten by the host and the young worms carried to different parts of the body in the blood stream. Some roundworms have special mouth-parts for burrowing into the skin and through the body of the host. Many roundworms that infect humans are quite small and harmless, but some can grow to an enormous size and may cause terrible diseases.

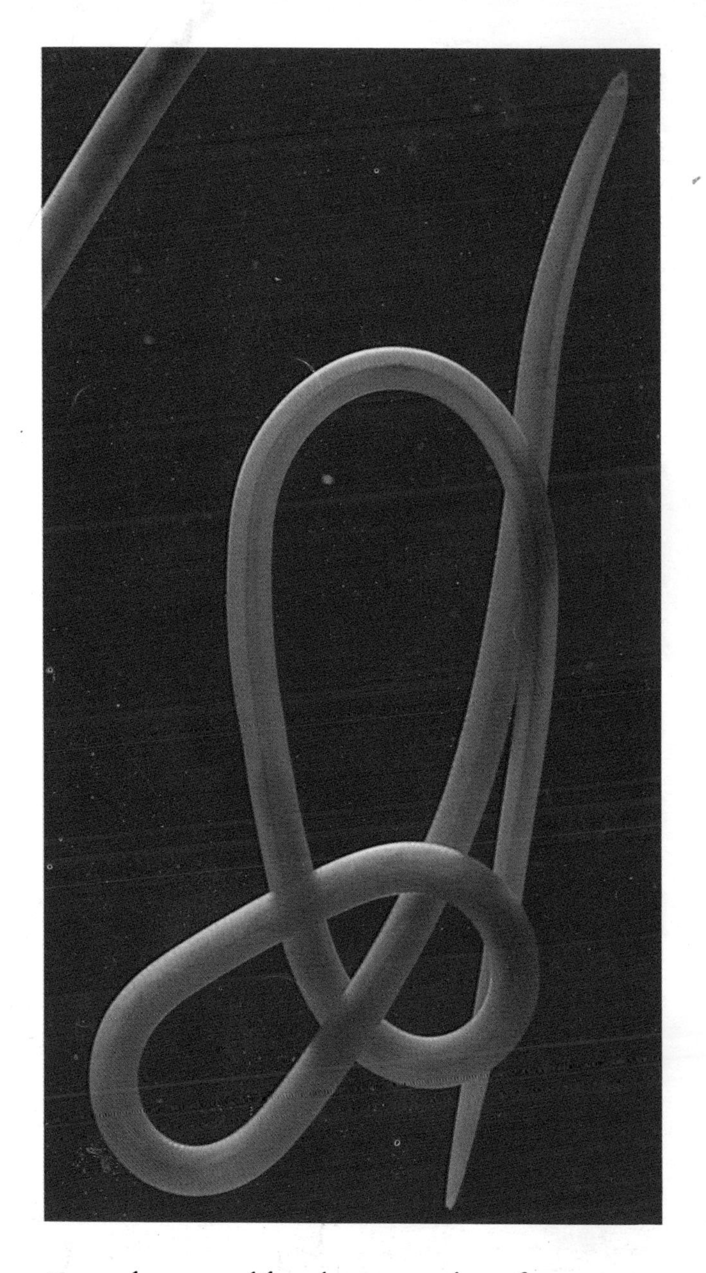

Roundworms, like this one taken from a pig, can grow as long as 30cm (12 inches).

False Worms

Some of the animals we call worms are not really worms at all. They may look very wormlike, but a closer look will show that their bodies are very different from real worms. Take a slowworm, for example. It has a long, smooth, wriggly body with no obvious legs. But it cannot be a worm because if you could look inside its body you would find it had a backbone. It is in fact a legless lizard and totally unrelated to any kind of worm.

Adult slowworms are 30 to 50 cm (12 to 20 inches) long.

This female glowworm gives out a luminous, yellow-green light from the end of her body to attract a mate.

The glowworm, or firefly, is another false worm. It is actually a beetle, and you can tell it is an insect by counting its legs. It has six legs, as all insects do.

Many other so-called worms are really the larvae or young stages of insects. Woodworms are the larvae of timber-boring beetles; mealworms are the larvae of flour beetles; and wireworms are the larvae of click beetles. Can you think of any other false worms?

How to Study Earthworms

Earthworms are fun and easy to study. Autumn is the best time to look for them especially on warm, damp evenings. Go out with a flashlight at night, or early in the morning, and you may catch them out on the surface feeding or mating.

Look for worm castings on the grass and do a count over a measured area. Look for other signs too, such as burrow entrances plugged with leaves, twigs or feathers. You might map out a piece of ground to study and keep records over several weeks or months.

You can find out more about earthworms and how they burrow by keeping them in a large glass jar or a tank called a terrarium. The pictures on these pages show you how to make a terrarium. If you fill it with layers of different colored soils you will be able to see how the worms mix up the soil.

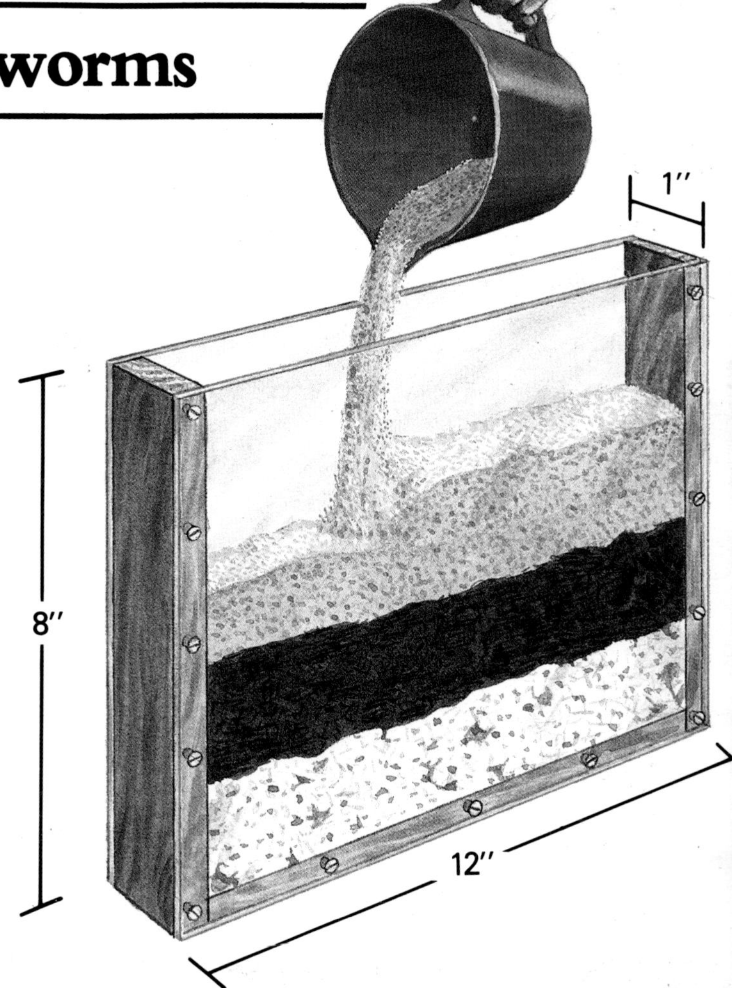

To make a terrarium, you will need 2 pieces of wood 8 inches by 1 inch, 1 piece 12 inches by 1 inch and 2 pieces of Lucite 8 inches by 12 inches. Glue the wood together to make a frame, then screw a piece of Lucite to each side of the frame.

Put different kinds of leaves on the surface and see which ones the worms pull down into the soil. Which end of a leaf do they grab? You can try feeding them with other things too — like chopped carrot, onion, raw meat or chocolate. Keep your terrarium in a dark, cool place and remember to water the soil occasionally.

Dig up some worms and take a close look at them. Try to count the number of segments — are there always the same number? Touch the body gently and see if you can feel the bristles. Place your worm on a damp piece of paper and watch it move. Never let a worm dry out, and put it back into the soil as soon as you have finished watching it.

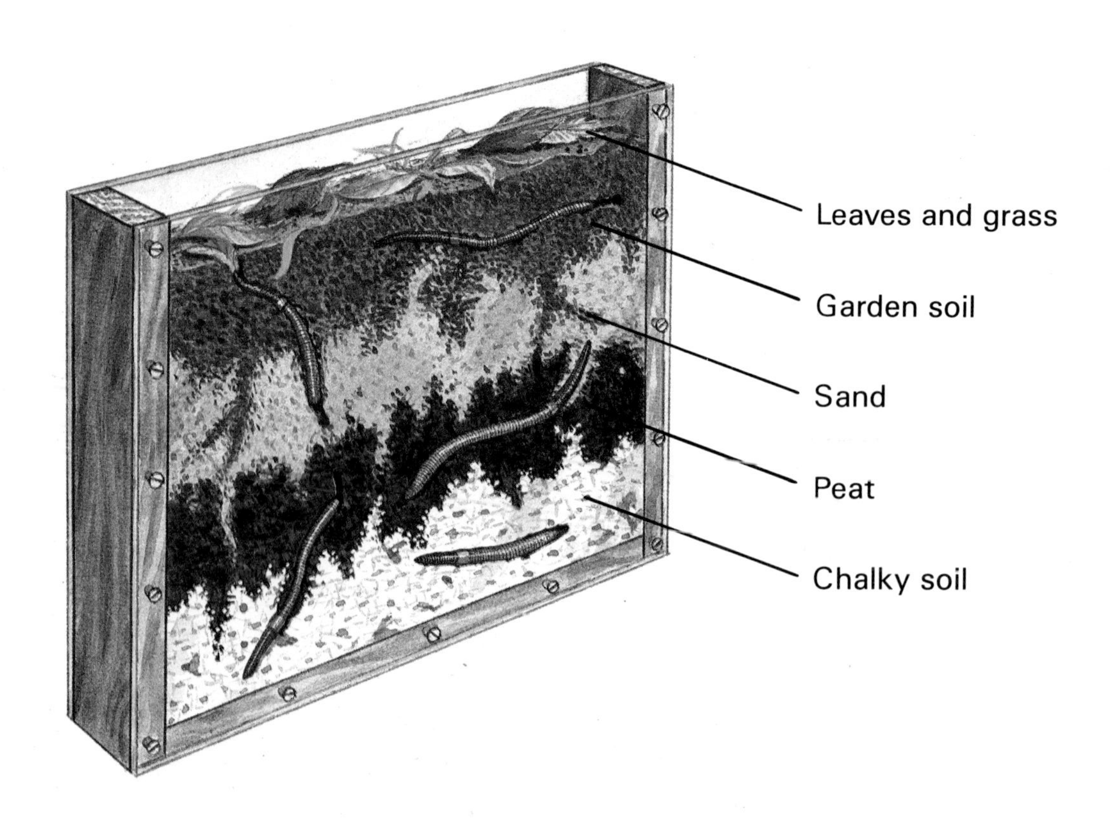

Glossary

Annelids A group of worms with segmented bodies.

Anus The opening of the gut at the rear end of the body, through which waste matter is passed out.

Carnivorous Flesh-eating; an animal that hunts or feeds on other animals.

Clitellum A thick band of skin, around the body of an adult earthworm or a leech, used during mating.

Gills The branched or featherlike structures used for breathing underwater.

Gizzard The part of a worm's stomach with strong muscular walls, for grinding up food.

Host An animal or plant in which a parasite lives.

Intestine The part of the gut from the stomach to the anus.

Larvae The young that hatch from the eggs of insects and some worms.

Mucus A slimy substance produced by the bodies of worms.

Parasite An animal or plant that lives and feeds on another.

Predator An animal that kills and eats other animals.

Prey An animal that is hunted and killed by another animal for food.

Proboscis The trunklike mouth-part of many flesh-eating worms.

Protein An important food substance, used by animals to grow or renew parts of the body.

Reproduce To produce more animals of the same kind.

Sperm The male sex cells used to fertilize a female's eggs.

Finding Out More

If you would like to find out more about worms, you could read the following books:

Buchsbaum, Ralph. *Animals Without Backbones: An Introduction to the Invertebrates.* Chicago: University of Chicago Press.

O'Hagan, Caroline. *It's Easy to Have a Worm Visit You.* New York: Lothrop, 1980.

Patent, Dorothy H. *The World of Worms.* New York: Holiday House, 1978.

Selsam, Millicent and Hunt, Joyce. *A First Look at Animals Without Backbones.* New York: Walker & Co., 1976.

Index

Picture Acknowledgments

Page 33 (top) by Dennis Green/Survival Anglia. All other photographs by Oxford Scientific Films by the following photographers: K. Atkinson 12 (bottom), 13, 28; G. I. Bernard 8, 12 (top), 14, 15, 18, 19, 22, 23, 31 (bottom), 32, 39, 40; R. Blythe 42 (bottom); J. A. L. Cooke cover, 21, 25, 26, 34, 35, 37, 41; S. Morris 38; J. Paling 27; P. Parks opp. title page, 9, 29, 36; D. Thompson 10, 11, 17, 24, 30, 31 (top), 33 (bottom), 42 (top). Artwork by Wendy Meadway.